# JAGUAR XK

## 120/140/150

# JAGUAR XK

## 120/140/150

Duncan Wherrett

First published in Great Britain in 1993
by Osprey, an imprint of Reed Consumer
Books Limited, Michelin House,
81 Fulham Road, London SW3 6RB
and Auckland, Melbourne, Singapore and
Toronto

ISBN 1855323788

Editor Shaun Barrington
Page design Paul Kime
Printed in Hong Kong

**Half title page**
*The leaping cat was a feature of the
Roadster and it became a general
factory option. The lights and chrome
fittings were as on the XK140. Fancy
saloon car gadgets such as door
handles appeared on the Roadster for
the first time*

**Title page**
*In April 1953 the third variation in the
XK120 range appeared in the form of
the Drop Head Coupe, this being a
much more practical convertible. Piping,
in leather cloth or PVC, ran along the
length of the rear wings*

**Front cover**
*An Old English White XK150 and the
famous Jaguar mascot; the leaping cat
came as standard on the 150 Roadster
– owners of other models could drive to
the factory and have it fitted
retrospectively or specify it as an option
before taking delivery*

**Back cover**
*First sold in Sweden, this rare (one of
only 199 built) 3.4S Jaguar XK150
Fixed Head Coupe was subsequently
re-imported into the UK and currently
resides in Sheffield, England, in the
hands of David Woodhouse*

For a catalogue of all books published by Osprey Automotive
please write to:

**The Marketing Department, Reed Consumer Books,
1st Floor, Michelin House, 81 Fulham Road, London SW3 6RB**

# Acknowledgements

I should like to thank the following owners and companies for their tremendous help and co-operation during the preparation of this Osprey Classic Marque volume. The time and assistance they provided during the photographic sessions was invaluable.

Roy Palmer, Hugh Palmer, Martin Morris, William Tuckett, Aubrey Funburgh, Classic Autos of Kings Langley, WP Automotive of Leatherhead, Jerry Stamper, Gavin Roberts, Rob Beere Engineering in Coventry, Roger Thorpe, Tony Locke, Brian Ekin, David Woodhouse, Alex Taylor and Aldridge Trimming of Wolverhampton.

Particular thanks are due to Michael Stewart, Patrick Lacey Restorations of Worksop and to J A Cook Coachworks of Nottingham for allowing repeated access to their cars and facilities.

# Contents

*With its relocated bulkhead and extended footwells, the Fixed Head Coupe was more roomy than the other models. The roofline was extended both forwards and rearwards by a total of 6.5 inches. Consequently, the doors were now nearly 6 inches wider and allowed access to the rear seats. The shorter front wings housed two 6 volt batteries. The XK140 FHC looked sleeker and more streamlined than its predecessor*

# XK120

The first Jaguar XK to be revealed to the general public was the alumium-bodied XK120 Open Two Seater (OTS), launched at the Earls Court Motor Show, London, in October 1948. Virtually bankrupt at the end of World War 2, post-war Britain had to endure a period of tremendous austerity; the rationing of even the most basic items, including food, clothing and fuel, remained a way of life. Suddenly, here was a new sports car of dramatic, luxurious appearance with racing car performance to match. The charismatic Jaguar was an immediate success, particularly in the United States, and the XK series remained hugely popular throughout the 1950s. Yet this fine new sports car was created somewhat by accident. Under the supervision of chief engineer William Heynes, Jaguar had developed the XK six-cylinder engine to power a planned high performance saloon. The 120 body was merely produced to wrap around the test-bed for the XK engine and it progressed from there.

Two models were announced at the 1948 Motor Show: the XK120 and XK100. The latter was to have the new four-cylinder, two- litre engine, developing 95 bhp. As the six-cylinder unit was proving to be far superior, only a few four-cylinder engines were made and the XK100 was never put into production. In contrast, the XK120 was a winner right from the start, its designation representing the car's top speed in miles per hour.

The first OTS, or Roadster, models were hand-built in aluminium over an ash wood frame in a manner reminiscent of pre-war coachbuilding methods. A total of 57 right-hand-drive and 183 left-hand-drive XK120 Roadsters were manufactured. Of these, none was actually sold in the UK. A handfull of XK120s were allocated to selected sources for competition use, while almost all of the remainder were exported. Only by agreeing to export at least 90 per cent of its output (a figure guaranteed to the government by Jaguar's managing director, William Lyons), was the company assured of receiving a sufficient allocation of Britain's then scarce metal resources.

As well as developing an enviable reputation for style and performance, the XK120 was also considered to be exceptionally good value for money, and the car proved to be very competitive in all markets. Once the position of the Roadster was well established, the range was expanded to include Fixed Head Coupe (FHC) and Drop Head Coupe (DHC) models.

*The aluminium XK120 as launched in October 1948. The styling of the car was very much the creation of Jaguar managing director William Lyons, who was knighted in 1956 and later became chairman of the company. With sleek lines, superb acceleration and a top speed of more than 125 mph, the model soon acquired a glamorous reputation and was billed as the world's fastest production sports car. This example is number nine off the Coventry production line and is finished in gold*

**Above**

*Originally called the XK120 Super Sports, the XK120 Roadster was strictly a two-seater with a removeable hood, sidescreens and windscreen. Although the body was aluminium, strength requirements dictated that the bulkhead, inner wings and part of the boot interior were made of steel. Production cars varied in some details (notably at the rear) from the original prototype*

**Left**

*Exciting open road performance was a salient feature of the XK120 and the car proved to be a fitting replacement for the pre-war SS Jaguars. In addition to its power and smoothness, the car was also praised for its excellent handling at speed. The 0-60 mph dash took around ten seconds. At Jabbeke, Belgium in May 1949, works test driver 'Soapy' Sutton recorded an officially timed 126 mph in an XK120 fitted with full weather equipment. The attendant party of journalists were even more impressed when 'Soapy' achieved almost 133 mph in the same car after it was 'cleaned up' by replacing the windscreen with a small cowl and adding a metal cover over the passenger compartment . In contrast, he ended this emphatic demonstration of speed and power by casually driving past his audience at a mere 10 mph in fourth gear, such was the XK120's low speed flexibility. Today, such performance would be hyped to the heavens, but, as California Autonews commented: 'It is typically British that Jaguars never claimed more than 120 mph for this car'*

**Left**
The powerhouse of the new XK range was the six-cylinder, 3442 cc unit, generating 160 bhp at 5000 rpm through a four-speed gearbox. It was destined to be used on a number of Jaguar models and to bring tremendous commercial and race success to the company for many years. It featured a high tensile alumium head with twin overhead camshafts driven by duplex roller chain. The carburettors were two, 1.75 inch SUs feeding combustion chambers designed by Harry Weslake, a master of gas flow theory; his contribution to the efficiency and power of the XK engine was immense. The early engines had cam covers with no studs at the front, as seen here, but oil leaks sometimes resulted. With the polished covers, the engine looked attractive as well as performing well. At this stage, the engine bay was very basic, comprising the engine, a grease gun, some

fuses and virtually nothing else. The four-speed gearbox had synchromesh on the top three gears. A close ratio gearbox and various different axle ratios became options

**Above**
The roof of the Roadster folded away behind the seats, with the side windows being stowed in the boot. There was no outside door handle on the Roadster – the side window included a flap to allow access to the interior handle. The support for the rear view mirror was quite short and the position of the rear window was such that with the roof raised, it was not possible to see through the rear window. Later windows were lower and, therefore, more useful. It was also possible to unzip the window and fold it down to improve ventilation

**Above**

*The aluminium models had more chrome than the later steel version and the interior mechanism of the roof was particularly smart. It is not easy to make such a roof which is effective against the wind and rain*

**Right**

*For what was really only meant to be a 'temporary' model, there was a tremendous attention to detail in the aluminium 120. The petrol cap, for example, incorporated three springs and a lock*

**Overleaf**

*The long, low lines of the XK120 Roadster. Although the later spats over the rear wheels were flat, at this stage that was not possible. The shape of the rear wings and panels necessitated that the spats were carefully made with a gentle curve in order to clear the rear hubs. At this time, the car weighed 22 cwt; the later steel-bodied car weighed an extra 3.5 cwt, which had a slightly detrimental effect on its performance. Brothers Roy and Hugh Palmer (who reside in Weybridge, Surrey, and Oakham, Leicester, respectively) imported two aluminium cars to England from New Zealand and Australia and they both required extensive restoration. Roy's silver car, serial number 26, was restored by Alan Holdaway and won the Jaguar Drivers' Club 'Champion of Champions' concours competition in 1991*

**Above**

*The side lights on the early cars were separate chrome fittings. In late 1952, the sidelights were welded on to the wing and in turn incorporated indicator flashers. At first there was no front flange on the fitting, so they were prone to lifting and developing corrosion. There were a number of subtle differences between the first aluminium cars and the later steel models. The light pods, for example, blended into the body in a more gentle way than on the later cars. The lines of the wings and the headlamps were also slightly different. Because the aluminium cars were all hand-built, there were small differences and discrepancies in the size and shape of some of the panels*

**Above right**

*Aluminium XK120s are so rare in Britain that the Palmer brothers had real problems in discovering exactly what was correct and how details should look. Their quest involved months of research, letter writing and telephone calls around the world. This boot shows the correct tyre attachment, jack and tool kit. The first 28 cars had the facility for hand-starting as well as electric, hence the starting handle*

**Right**

*Not just the early cars but many of the later ones also had aluminium boot lids, bonnets and sometimes doors. Part of the ash frame can be seen here. The centre space is used to accommodate the spare tyre but, to be strictly correct, the side sections should be enclosed*

**Above**

*These large rubber grommets are unique to the aluminium cars and help to cushion the strain of the window frame on the metal. Roadster windscreens could be removed and replaced with small racing screens*

**Left**

*The seats in the XK120 had rather flat backs and were not ideal when taking bends at any sort of speed. Once the C-type came on the market, the seats from that model became an optional extra and offered considerably improved lateral support. The large steering wheel gave the driver little space around the legs. Roadsters came with a tonneau designed to cover the whole cabin or unzip to allow driver access*

**Left**

*The dash of the Roadster was covered in leather. Instrumentation was laid out clearly and logically, displaying essential information as well as incorporating a thermometer and clock. By pressing a button, the gauge for showing how much fuel is in the petrol tank was also designed to show the oil level*

**Above**

*The doors were opened with the aid of a cord pulley; also inside the door was a flap covered pocket. The early cars had no door stops so care was required when opening the doors in order to avoid damaging the side panels. The two chrome knobs are for attaching to the side screens. On the steel Roadsters, the lower door panel was covered in leather*

**Above**

*The steel-bodied XK120 was introduced in April 1950, after which production of the car began to increase substantially. Indeed, the Roadster was destined to become the most popular model in the XK120 range, with more than 7600 being produced between 1949-1954. Described by Jaguar as the XK120 Open Two Seater Super Sports, the car was priced at £1263*

**Left**

*An original brochure for the XK120 Roadster. The high relief bonnet badge was particularly attractive. The grille is slightly different from those fitted to later cars in that there is an extra ridge down the front of each strip. A technical drawing of the original prototype which appeared in the same brochure included domed headlights, spats for the front wheels and a racing-style streamlined headrest for the driver, items that were not incorporated on production cars*

**Above**
The XK120 chassis was based on the Jaguar Mk V saloon's, although it was much narrower. The side members are of box section and rise up over the rear axle, which allows the floor to be lower

**Left**
Judges considering the finer points of an XK120 at the Jaguar Drivers' Club 'XK Day' at Stamford, Leicester, in 1992. The event attracted a glittering array of Roadsters, Fixed Heads and Drop Head Coupes from across the range

**Above**

*Rear suspension was by means of seven, long semi-elliptical springs and Girling lever-type shock absorbers on a solid rear axle. Lockheed hydraulic brakes with 12 inch drums were fitted all round; under intense race conditions they were prone to fading*

**Left**

*The front suspension was independent, featuring upper and lower wishbones with ball joints on the stub axles, telescopic springs and dampers. The upper wishbone is attached to a bracket which also connects to the steering mechanism. This gives extra support against lateral wheel movements. Springing was assisted by longitudinal torsion bars, running from the lower wishbone to the central chassis area. A rubber mounted anti-roll bar is also fitted on the front*

**Above**

*The interior of the FHC gave the vehicle more of a saloon car appearance, being well trimmed throughout in leather, wood, quality cloth and incorporating courtesy lights. Behind the seats were a useful shelf and luggage box. The impressive performance of the XK was now matched to saloon comfort*

**Right**

*The XK120 Fixed Head Coupe was launched in March 1951, offering year-round weather protection for the car's occupants – especially useful in some of the world's more unsettled climates. Doors are slightly wider and footwell vents are fitted. Special Equipment models became available in 1952. Wire wheels were an option, although the spats could no longer be fitted. Power was up-rated to 180 bhp matched to a Burgess straight-through silencer. Stiffer front torsion bars and seven leaf rear springs were fitted to improve the handling. Self-adjusting front brakes were introduced in April 1952. It was in a similar XK120 FHC that a unique motoring record was established in August 1952 at the Montlhery banked circuit near Paris. Racing driver Leslie Johnson instigated an attempt to average 100 mph for seven days and nights. The Jaguar was virtually standard except for some optional extras such as highlift cams, an 8:1 compression ratio, twin exhaust, knock-off wire wheels and high final drive ratio. As the days past, Johnson, together with Stirling Moss, Jack Fairman and Bert Hadley, drove the car to four new world records, as well as Class C records. Unfortunately, a broken rear spring ended any further official record attempts, as only spares carried in the car were allowed. Nevertheless, the drivers pressed on. After covering 16,851.73 miles in seven days and seven nights, the team averaged 100.31 mph – the first time such sustained speed performance had been achieved. The record-breaking XK120 is today preserved as part of the collection owned by Jaguar Cars at their Coventry factory in the West Midlands*

**Above**

Costing £1660, the Drop Head Coupe was announced in March 1951 and featured a lined folding hood; top speed was slightly reduced at 119.5 mph and 0-60 took 9.5 seconds. A total of 1765 Drop Heads were built before production switched to the XK140 in late 1954. Unlike the Roadster, the DHC had wind-up windows – and useful quarterlights

**Right**

The roof of the DHC folded down neatly, and tucked away under a separate cover. This example, restored by Michael Stewart of Macclesfield, Cheshire, is in pastel green with a suede green interior. Customers unhappy with Jaguar's standard colour schemes could specify any colours on special order and at extra cost. The hooks on top of the windscreen for attachment to the hood are adjustable to give the correct tension; this refinement was not available on the aluminium-bodied model

**Above**

*Attractive yet functional, the rear light had a busy time, working as rear light, brake light and indicators*

**Right**

*The DHC transformed the XK120 into a sports car that was ideally suited to long distance touring – the car had plenty of power for relaxed high speed cruising and a comfortable cabin to keep tiredness at bay. Air vents are let into the sides to ventilate the footwells, a modification introduced in late 1951*

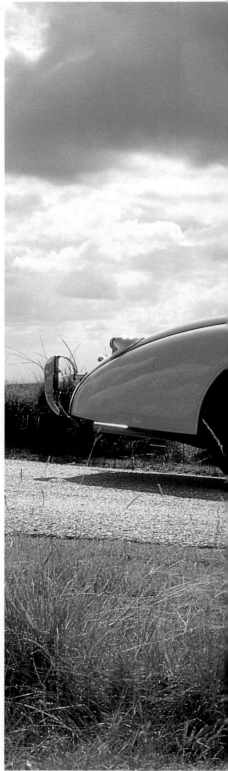

**Above**

*The rear of the XK120 was simplicity itself and displayed only the bare essentials. There were no real bumpers, only overriders whose supports were usually painted black, although on the aluminium model they were in body colour*

**Right**

*The front bumpers were similarly light in structure. Under the front bumpers were air intakes which help to cool the brakes and lower parts of the engine. The original number plates were made of high relief plastic. Standard wheels were pressed steel discs, finished in body colour. Wire wheels came as a Special Equipment item in chrome, silver or body colour. Other SE components included high lift cams, a 180 bhp engine, stiffer torsion bars, stiffer rear springs and two fog lamps*

As with the fixed head model, the dashboard was of walnut veneer. The handbrake was of the racing 'fly-off' type. The steering wheel (white on most export models) was adjustable for distance and a factory-fitted radio was an optional extra

**Above**
*Two six-volt batteries were fitted behind the seats and linked to power the 12 volt electrics*

**Right**
*Part of the engine bay of an XK120 during its restoration. Visible on the bulkhead is the switch for the heater, the motor for the windscreen wipers, the brake reservoir and oil and petrol pipes*

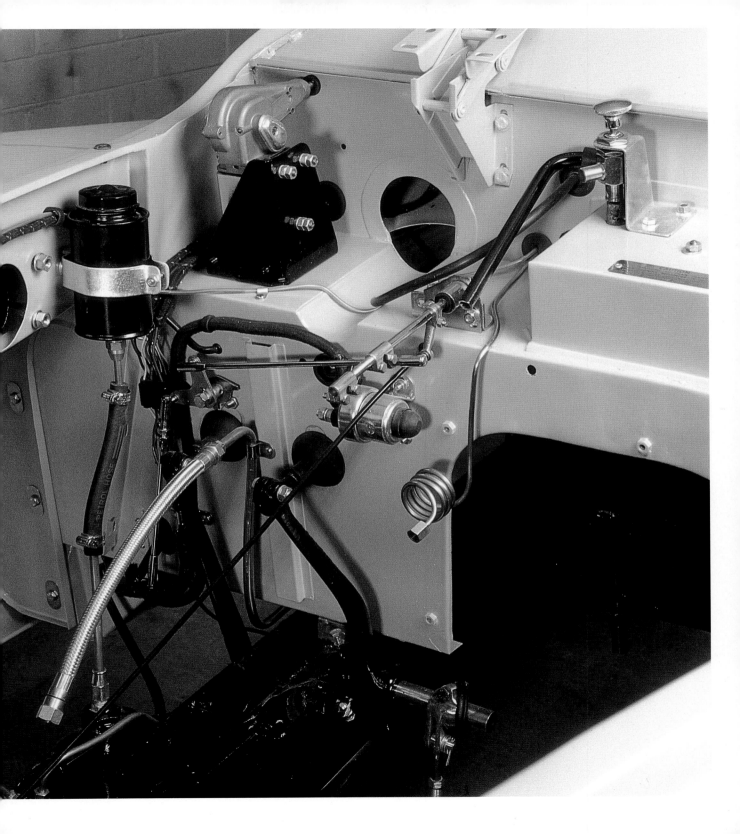

**Above**

The bonnet of the XK120 was long and flexible, in spite of reinforcements at the front, middle and the rear. Restorers often find it difficult to make the bonnet fit properly — with use it often fails to stay flush with the engine bay sides. Early bonnet hinges tended to be rather weak and were soon strengthened

**Above right**

The XK engine changed little throughout the range. The plug leads were neatly held in place and the grease gun was part of the original equipment. There is often controversy about such things as to whether a bonnet support should be chromed or wheel arches should be black or in body colour. In fact, the specification of such items varied at times and what is correct will depend on the car's number and its precise date of manufacture

**Right**

One owner's modifications to a 120 dash, incorporating turbo controls and extra fuel tank switches

# Racing models

The XK120 was ideally suited to the competitive arena. Its high power-to-weight ratio, predictable handling and good streamlining made the car immediately at home on the track. Early examples, in virtually standard trim, were raced successfully around Europe, North America, Australia and New Zealand.

One of the more notable XK120s was the rally car driven by Ian Appleyard, with his wife Patricia (the daughter of Jaguar supremo William Lyons), as co-driver and navigator. The Appleyards proved to be a very effective combination during the Alpine, Coupe des Alpes and RAC rallies between 1950-52.

The year 1950 also saw three privately owned and virtually standard XK120s competing in the Le Mans 24 Hours. Two of the Jaguars crossed the finishing line in 12th and 15th place, in the circumstances an extremely creditable performance. Not surprisingly, William Lyons was soon persuaded by Jaguar's chief engineer Bill Heyes and F R W 'Lofty' England to authorise the preparation of a car specifically for the 1951 Le Mans event.

In spite of its successes, the standard XK120 was neither light enough nor strong enough to cope with sustained racing speeds approaching 150 mph. Jaguar's Malcolm Sayer drew on his experience as an aircraft designer to create a new lightweight, aerodynamic body in magnesium alloy which was welded to a tubular frame; the result was the XK120C or C-type.

As Le Mans loomed rapidly on the horizon, Jaguar decided to prepare three modified XK120s as an insurance policy in case the C-types were not completed in time for the race, but these were not needed.

The C-types were privately entered so as to minimise any possible embarrassment to Jaguar if the cars performed poorly or failed to finish. After actually being *driven over* to Le Mans, and despite never having been raced before, the C-types exceeded all expectations. Peter Walker and his co-driver Peter Whitehead won the race, their C-type averaging 93.49 mph. Jaguar were rampant, Stirling Moss setting a new lap record of 105 mph. The world-wide publicity and sales potential from such a win was considerable.

In 1952 Jaguar embarked on a full race programme, but a repeat of their success at Le Mans eluded them. In an effort to maintain the competitive edge of the C-type, Jaguar redesigned the front of the body to improve the car's streamlining. The C-type's new, longer nose certainly made the car look faster, but it had been built with too few cooling vents and, inevitably, all three Jaguars were forced to retire from the race with overheating problems.

It was a different story in 1953 when Jaguar returned to Le Mans with three improved and lighter C-types. The body shape was similar to that fitted to the 1951 cars, and the XK engine was given triple Weber carbs and other detail modifications for improved performance. Duncan Hamilton and co-driver Tony

*This lightweight C-type, seen in company with a contemporary XK120 Roadster, was the works car driven to fourth place by Peter Whitehead and John Stewart in the 1953 Le Mans 24 Hours. At Jabbeke, Belgium, in October 1953, Norman Dewis, Jaguar's chief tester, achieved an officially observed and timed 178.3 mph in a C-type; even more remarkably, he had earlier attained 172.4 mph in a specially modified XK120! The C-type's disc brakes were a first at Le Mans and enabled it to consistently out-brake the opposition. With triple Weber carbs, racing pistons, 8:1 compression ratio, high-lift cams, reinforced crankshaft and twin exhausts, the engine developed 210 bhp at 5800 rpm. The cooling system was improved by fitting a more efficient water pump and an aluminium radiator. Firmer suspension and changes to the torsion bars increased stability during cornering and acceleration. At the end of 1953, the car was sold to Ecurie Ecosse, where it enjoyed considerable success over the next two years*

Rolt emerged victorious, taking the chequered flag after averaging 105.95 mph – the first 100 mph-plus win at Le Mans. Jaguars also came second and fourth.

The D-type was originally built for Jaguar's continuing campaign at Le Mans and was not meant to be a road-going sports car. At the D-type's first appearance at Le Mans in 1954 it was troubled by misfiring problems (eventually traced to blocked filters, though not before much valuable time had been lost); after a storming drive, the D-type of Rolt/Hamilton managed a magnificent second place, victory narrowly going to Gonzales' 4.9 litre Ferrari. A month later at the 12 hour event at Rheims, the Jaguars took first and second places after Stirling Moss had set the pace (and broken the Ferraris) with a drive of blistering power.

A works 'long nose' D-type driven by Mike Hawthorn and Ivor Bueb won at Le Mans in 1955, but only after the leading Mercedes 300 SLR piloted by Juan Manuel Fangio and Stirling Moss was withdrawn on orders from Stuttgart

**Above**

*The whole of the tail section is occupied by a 40 Imp gal fuel tank. Individual lights were fitted to illuminate the car's number during the Le Mans 24 Hours. The highly geared rack and pinion steering mechanism was similar to that fitted to the later XK140 series*

**Right**

*A rugged day for a rugged car. The C-type's startling, streamlined curves were the creation of Malcolm Sayer, whose imaginative use of aerodynamic theory and monocoque construction techniques came from his work with the Bristol Aeroplane Company in World War 2. Sayer was also responsible for the body of the later D-type and, most notably, the legendary E-type – the only production Jaguar not designed by Sir William Lyons. This C-type, resplendent in British Racing Green, has been in private hands since it left the factory in the mid fifties. In 1970, the car was purchased complete (but in pieces) by Aubrey Finburgh of Classic Autos in Kings Langley, Hertfordshire. Since being restored, the C-type has been a frequent competitor in historic racing events at Spa, Belgium, and Monterey, France, as well as various continental tours and hill climbs – regardless of weather conditions*

**Left**

The modified XK engine was supported in a rigid lightweight tubular frame. Two inch steel tubing with 1 inch and 1.5 inch diagonal bracing ensured that the car's box frame was both light and strong. Although there was a little stressed panelling, the car's strength was almost entirely in the frame, part of which may be seen here. The front suspension of two wishbones, telescopic dampers and torsion bars was similar to the XK120s. Small doors were fitted so as to avoid weakening the bodywork. A spare set of spark plugs was stored to the right of the steering wheel. These seats have been covered in the original type of material, which was notoriously good at absorbing rain water

**Above**

A two inch carburettor on a 'C' head as fitted on the production C-types. The 1951 Le Mans cars had twin one and three quarter inch SU carbs, while the 1953 cars ran with triple Webers

following the horrific crash of Pierre Levegh's Mercedes, which not only killed him outright but also claimed the lives of 80 spectators.

The 1956 Le Mans event turned into something of a fiasco for the works Jaguars. On only lap two, Paul Frere decided to overtake team mate Jack Fairman as their D-types approached the Esses; Frere applied his brakes too late on the damp track and spun after hitting the barrier. Fairman also spun (deliberately) to miss the impetuous Frere; unluckily he was rammed by another errant competitor.

As if that wasn't enough, the surviving Hawthorn/Bueb car suffered from repeated misfiring. The culprit turned out to be a cracked fuel line, but by the time this was discovered the car had slipped out of contention. With the consolation of establishing the fastest lap as they battled through the field, Hawthorn/Bueb came home in sixth place. But Jaguar's blushes were more than spared by the single exploratory entry of Ecurie Ecosse, an ex-works D-type driven by Ninian Sanderson and Ron Flockhart. Amazingly, the team triumphed at their first attempt.

After Jaguar had announced their retirement from racing in October 1956, Ecurie Ecosse took up the challenge at Le Mans the following year with four factory prepared D-types. The team took first, second, third and fourth places. Clearly, the D-type was in a league of its own.

One of the great virtues of the C- and D-types was their outstanding reliability, but in the late 1950s international races were restricted to three litres. At this capacity, the D-types did not prove to be as reliable in long distance events and so the Jaguar factory no longer entered the car competitively; subsequent entries were all pirate.

*Designer Malcolm Sayer went back to the wind tunnel to optimise the shape of the D-type, and made more extensive use of aircraft construction methods. This D-type, regularly raced by William Tuckett, is a 'short nose' production model – only the factory works cars had the fin on the back. In its heyday, the car was raced by the great Jim Clark and, driven by Henry Taylor, ran third at Spa in 1957*

**Above**

*The 'aircraft on four wheels' had an excellent co-efficient of drag, its low frontal area being the result of tilting the engine forwards at an angle of eight degrees and adopting dry sump lubrication, thereby reducing engine height. As a result the lowered body also had beneficial effects on the car's centre of gravity. Despite this the D-type had a reputation for being somewhat skittish in the wet, but there is no doubt that it handled well in the right hands*

**Left**

*The magnificent XK engine could hardly be improved upon, cubic capacity remaining at 3442 cc; compression ratio was 9:1, while triple Weber carbs boosted power to 250 bhp at 6000 rpm and generated 242 lb/ft torque at 4000 rpm. The engine's block and cylinder head were identical to those fitted to the production road cars. The coolant header tank may be seen above the radiator. An additional oil pump and oil cooler are fitted, baffles and a breather pipe in the oil tank preventing aeration in the lubricant as it circulated at high pressure. A large triple dry-plate clutch and a crankshaft torsional vibration damper replaced the flywheel*

The aerodynamic body of the D-type was deceptively compact, overall length being 12 ft 10 in. Under the skin the new car was rather different to the preceding C-type. Although tubular framing was retained to support the engine and front suspension, the car was mostly of monocoque construction. The immensely strong central tub was made of magnesium alloy and incorporated double-skinned front and rear bulkheads and substantial, tube-like sills. The aluminium body panels were rivetted into place. The rear bodywork, holding the flexible fuel tanks and spare wheel, as well as the aft trailing link suspension units, was bolted on to the tub at the rear bulkhead and could be removed when necessary. As on the C-type, the one-piece bonnet hinged forwards. Incidentally, the provision of a spare wheel was obligatory at Le Mans

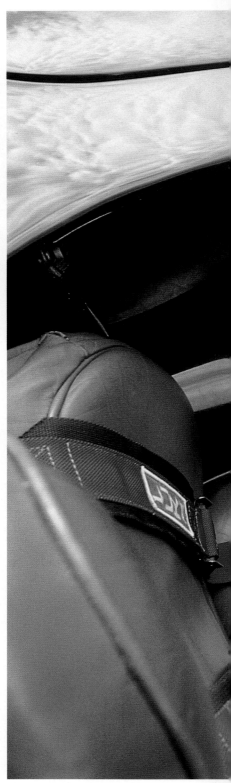

**Above**

*Discussion time before the Duncan Hamilton Trophy race at Oulton Park in Cheshire during August 1992. The works D-types all had fins to improve directional stability at speeds approaching 200 mph. Considering that both the C and D-types were designed and built specifically for the fast Le Mans circuit with its long Mulsanne Straight, the cars proved to be remarkably competitive at a variety of tracks in Britain, Europe and the United States. Apart from the ex-works cars raced by factory drivers, 'production D-types' were supplied to privateers such as Briggs Cunningham (with whom drivers such as Walt Hansgen achieved notable success in the States between 1955-58), and the famed Scots team of Ecurie Ecosse, whose drivers included Jimmy Stewart, elder brother of the future F1 world champion, Jackie*

**Right**

*Cockpit facilities were kept to a minimum and consisted of a standard light alloy, wooden-rimmed racing steering wheel, basic engine information dials and a padded head restraint. The 1956 works cars had a longer nose and compulsory wrap-around windscreen*

**Above**

This array of classic cats includes an XKSS, D-type, C-type, and XK120 racer. Perhaps surprisingly, production of the C and D-types totalled a mere 53 and 71 respectively

**Left**

The rear drive unit was similar to that found on the road cars, but a higher final drive ratio of 2.79:1 and a close ratio gearbox enabled the D-type to attain 180 mph at 6000 rpm; 100 mph could be reached in under 13 seconds. Dunlop disc brakes were retained, 12.75 inch discs having three pairs of pads on the front and two on the back. The front discs were ventilated by air scoops. The Frank Sytner car seen here is fitted with the later 3781 cc engine

**Above**

Jaguar supplied their XK engines and components to a number of specialist racing car builders such as Cooper, Tojeiro and HWM, but their best customer was the Cambridge-based engineering firm of George Lister & Sons. The brainchild of Brian Lister and driven by redoubtable Archie Scott-Brown, the first Lister-Jaguar took to the track in 1957 and went on to win 11 races from 14 starts. The following year the company combined the XK engine and drivetrain with a chassis and low-drag body of their own design. Later dubbed 'knobbly' Listers due to their distinctive shape, the cars were rarely beaten in club and national events around the world before taking well earned retirement in late 1959. Lister-Jaguars were particularly dominant in the United States, and excelled on the relatively slow, tight circuits where the standard D-types and big Ferraris were at a disadvantage. Today, Steve O'Rourke is a regular competitor in historic events with his much-prized genuine Lister 'knobbly'

**Left**

The XKSS was developed as a road car from the D-type – largely as a means of exploiting spare D-type monocoques. There was a particular demand in the United States for a car with racing performance and some degree of touring capability which could participate in American Group C production races. Unfortunately, a fire at Jaguar's Coventry factory in February 1957 halted production before the number of cars required for Group C qualification had been manufactured. The XKSS was powered by the tuned 3442 cc engine and fitted with Dunlop disc brakes; the frame and body panels were identical to the D-type. Performance was nearly as quick as the D-type: the 0-60 dash took a mere 5.2 seconds and 100 mph followed in under 14 seconds. Inexplicably, a few of the 16 cars built have been converted to D-types, which means they end up neither a genuine 'D' nor an XKSS

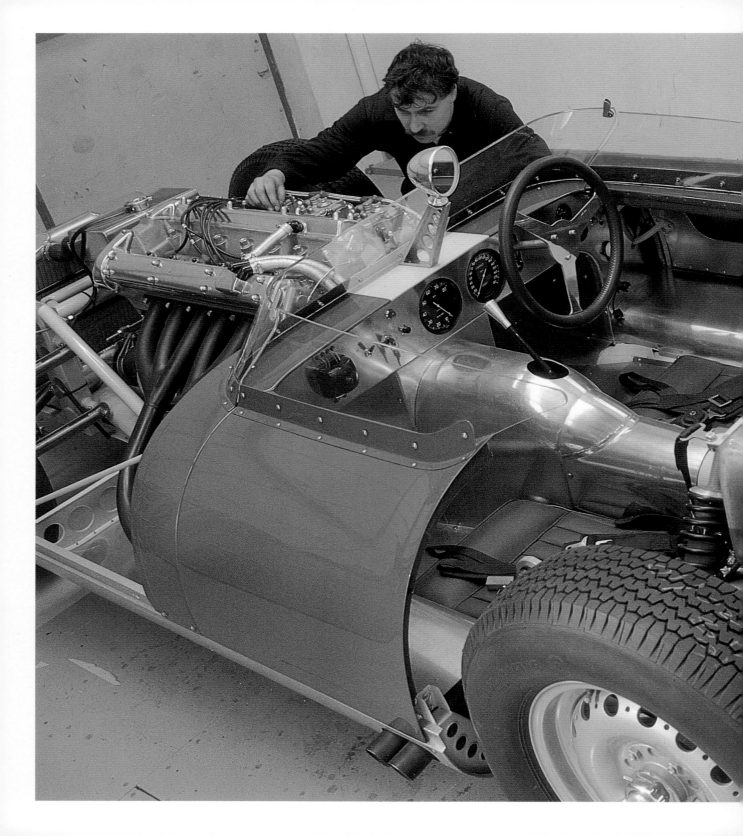

**Above**
*Detail of the radiator and oil cooler of the Lister 'knobbly'*

**Left**
*In 1985, 12 new 'knobblies' were built to mark the centenary of George Lister & Sons. Using the original jigs and drawings for the cradle-style frame and three of the original workers, the chassis were made at the Lister works, the bodywork being sub-contracted to Shapecraft. Supervised by Jaguar high performance specialist Lawrence Pearce, the remainder of the production work and final assembly was carried out by WP Automotive of Leatherhead, Surrey. The quality of the engineering and workmanship involved is absolutely first class and every effort has been made to faithfully replicate the most incidental detail. Wisely, however, the new 'knobblies' are fitted with a handbrake, absent from the originals in order to save weight. The 3781 cc engine with wide angle head and dry sump lubrication generates 300 bhp and 225 lb/ft torque at 5000 rpm. As before, the 'knobbly' doesn't hang about, 0-100 mph being achieved in 11.2 seconds. The cars are finished in the authentic green and yellow colours of former sponsor British Petroleum*

**Overleaf**
*Replica Jaguars abound – from XK120s in glassfibre to D-types in aluminium. The quality of such reproductions varies considerably, but they do offer many people who are unable to afford the genuine article the chance to drive a 'Jag-wah'. This fine example is a replica Lister 'knobbly' owned by Eike Wellhausen from Chesterfield, NE Derbyshire*

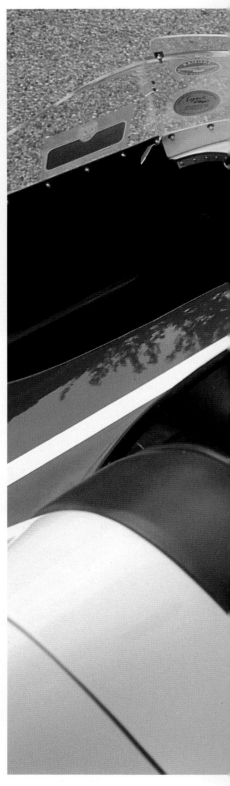

**Above**

In the late 1950s, Briggs Cunningham commissioned the building of a Jaguar for a private entry in the Le Mans 24 Hours. The 2997 cc engine featured new light alloy casings, plus a modified head and fuel injection. The chassis was similar to that of the D-type but was extended to support the independent rear suspension and boot. The rear disc brakes were now inboard and attached to the halfshafts with universal joints. Special cooling was developed for the rear brakes and the differential. Finished in flag blue and early old English white, this 1990 replica of Cunningham's car was made by Lynx for Jerry Stamper. Based at St Leonards-on-Sea in Sussex, Lynx replicas are among the best, being handcrafted with aluminium panels

**Right**

The 'long nose' D-types of the time came with a rear exhaust. However, Briggs Cunningham wanted a side exhaust, so that is what he (and Stamper's replica) got. Nothing, it seems, is too much trouble for Lynx – having learned that the original car's wrap-around windscreen was changed, they even reproduced the left-over rivets ...

# XK140

By 1954, the general appearance of the XK120 had become somewhat dated and the face-lift came in the form of the XK140. It was also necessary to have more space for people and luggage. Accordingly, the engine was moved forward by three inches and with it the scuttle and windscreen, but in addition to creating extra legroom, this also had the added benefit of improving the front-to-rear weight distribution and handling. Considerably lighter in operation, rack and pinion steering superseded the Burman recirculating-ball mechanism of the 120s. The changes were not too dramatic, so the marque was able to attract new enthusiasts without losing the existing ones.

As before, three body styles were offered (Open Two-Seater, Fixed Head Coupe and Drop Head Coupe), all of the cars being released in October 1954. The XK140s were fitted with the former Special Equipment engine developing 190 bhp at 5500 rpm. Compression ratio was 8:1 with high lift cams; the engine had steel, rather than cast iron, main bearing caps to cope with the power increase, and an eight blade fan, larger radiator and better water pump improved cooling. To take good advantage of the extra horses, the gearbox had closer ratios. Laycock de Normanville overdrive became an option and, if matched with a lower final drive ratio, the car could cruise comfortably at 100 mph. The XK140 Special Equipment model could reach 130 mph. Automatic transmission – vital for the US market – could also be specified from early 1955.

Between 1954 and 1957, XK140 production amounted to 3354 Roadsters, 2808 Fixed Heads and 2789 Drop Heads. The vast majority were exported to the USA; of the 73 right-hand drive Roadsters, only 47 stayed in the UK. The number of rhd Fixed Heads and Drop Heads was 843 and 479 respectively.

**Left**

*For the XK140 Special Equipment the engine delivered 210 bhp with the C-type cylinder head and dual exhaust. The SE featured wire wheels and fog lamps. Titled XK140M in the United States (XK140MC with the C-type head), the SE model offered extra performance and refinement*

**Above**

The overall lines of the XK140 models were similar to the 120s, those of the Roadster being much the same as its predecessor but with additional chrome trim. Resplendent in Carmen red, this Roadster represents the marque most effectively. Classic Jaguars seem to able to park wherever they like ...

**Right**

The interior is virtually the same as on the XK120, with a dash covered in leather and the same door trim. The instruments and switches remained grouped in the centre, making the dash suitable for both right and left-hand drive cars. With the raised bulkhead and a universal joint on the steering column, the wheel was now slightly higher and its reduced diameter made the driver less cramped in his seat. As well as making the steering lighter, the change to rack and pinion (with rubber mounting on the chassis to absorb shocks), also gave the car a more responsive feel. The XK140 Roadster is a rare car in Britain and this machine was imported from the United States by Gavin Roberts. Although converted to right-hand drive, the position of the handbrake is unchanged

The straight six – the backbone of the XK's tremendous power. With its standard crank, pistons and valves, the original engine requires little work for normal use. This crankshaft has been fitted with a lip seal conversion by Rob Beere Engineering of Coventry, thereby eliminating the usual oil leak from the flywheel end. The seven bearing fully counter-balanced crankshaft was enclosed in a heavily webbed crankcase, its strength no doubt contributing to the engine's smooth running. The oil pump delivered lubricant to each main bearing through channels running the length of the crankcase – this maintained good oil flow at low velocity and eliminated frothing. Such a feature was significant, as an early prerequisite for the engine was longevity. Standard compression ratio was a modest 7:1 on 72 octane fuel. A high pressure water pump ensured a constant temperature around the inlet manifold, across the head and around each cylinder

*Perhaps many people's favourite XK140 is the Drop Head Coupe. Along with the increased engine power, the car incorporated some suspension modifications. The rear units featured Girling telescopic shock-absorbers, with larger diameter torsion bars (from the XK120SE) being fitted on the front*

**Left**
*An evolutionary design, the body line of the new car had the same styling as the XK120. There were, however, a number changes to the exterior; immediately apparent were the much heavier front and rear bumpers with overriders. Over the years, most owners have fitted wire wheels, although originally these only appeared on SE models or as an optional extra. The pressed steel wheels had chrome hub caps, and spats were retained over the rear wheel arches. Without the spats, the wheel arch was edged with a strip of brass beading. As restored by Michael Stewart, the body colour of this XK140 is in battleship grey*

**Above**
*The tail light was now much more substantial, integrating the rear light, brake light, indicator and a reflector. The number plate was relocated on the fixed panel below the now lockable bootlid*

**Previous page**
*The interior was as luxurious as ever, with a figured walnut dashboard, wood trim on the doors and leather upholstery throughout*

**Above**
*The XK140 DHC was everything one expected of a sports touring car. The wheelbase was unchanged, but the car was stretched by three inches so as to increase the size of the cabin and permit the incorporation of small rear seats, although these were only really suitable for children. No longer found behind the front seats, the two 6 volt units were replaced by a single 12 volt battery placed behind the left front wheel*

**Right**
*With the heavier bumper and overriders, the front of the car had a much more solid appearance. Flowing lines still predominated, however, and there was little buffeting from the wind at speed. Though similar in size and shape to that on the XK120, the grille was now cast in one piece as a cost-saving measure with seven thicker blades rather than the previous 13. The bonnet badge was integrated with the grille. Chrome trim strips were placed on the bonnet, and flashing indicator lights were fitted on the front wings*

**Left**

*There was more chrome on the boot, which had a new boot handle and a new locking system. Due to a punctuation error, the commemorative badge on the boot handle implied that Jaguar won the Le Mans 24 Hours on three consecutive occasions, whereas in fact the victories occured in 1951 and 1953. A combined reversing/number plate light completes the ensemble*

**Above**

*It was necessary to turn on this switch under the bonnet before the heat could be directed into the cabin*

**Above**
*One of the neat details of the DHC was this catch for holding the hood in place when folded down*

**Right**
*A luggage rack and picnic basket was often de rigueur in the summer*

**Above**

*Jaguar rallies and shows are always well attended. For many enthusiasts, having an entertaining day is equally as important as the result of the Concours*

**Right**

*The essence of XK motoring: Roger Thorpe drives through open country in his FHC*

**Above**

*Numerous options were available on special order for XK140 customers, including wire wheels, windscreen washers, fog lamps, and the C-type head. Apart from the early examples, the 'C' heads were identified with an embossed badge, a 'C' in the centre and red paint. This engine has been modified with HD8 E-type carbs and will deliver 225 bhp*

**Left**

*A front end which really looks as though it means business. The bonnet strap is not purely for decoration, however. The bonnet is held in place by just one small spring and it is not uncommon for a catch to release itself, causing the bonnet to fly up uninvited at speed. A small strap attached to the bottom of the grille is a common alternative. The vents are useful for cooling if high revs are maintained for long periods. The leaping cat only became available on the later XK150 Roadster, but after that time the factory would fit it on other models as an extra; many owners did the job themselves*

**Above**
The larger Le Mans filler cap has no functional advantage because the petrol pipe is of the same diameter as the standard fitting – but it looks good

**Left**
The revised steering wheel no longer had the large protruding central horn and the indicator switch for the FHC and DHC was placed on top of the dash. With the bulkhead moved forward, the overall increase in the length of the cabin was about five inches. The roof on the XK140 FHC was one inch higher and, together with the larger interior and greater range of adjustment for the front seats, made the car more comfortable – particularly for taller drivers. The XK140's firmer ride and more communicative steering gave the driver a greater feeling of control. With the batteries now in the front and the engine moved forward by three inches, the weight distribution was changed for the better. Together with the strengthened suspension, cornering characteristics and general stability were significantly improved. Tony Locke has rebuilt his XK with international rallies and historic racing in mind

**Above left**
The rear end looks rather like a heavy and solid saloon car, which belies the XK's speed and sporty character. Although the large bumpers fitted to the car attracted a fair amount of criticism, they were undoubtedly strong. The bootlid of the 140 series was six inches shorter, ending above the rear bumper

**Left**
As displayed by this XK140 Roadster, factory options included bucket seats, racing clutch and tyres, 9:1 compression ratio and racing-style windscreens

**Above**
The XK's are popular rally cars and this model is taking a gentle turn around the Donington Park circuit in Leicestershire

At a time when some owners are too frightened even to take their cars out of the garage, it is a rare delight to see an XK140 on the track. As he demonstrates here, Tim Kemp is no slouch when it comes to hill climbing

# XK150

The XK150 Fixed Head and Drop Head Coupes were launched in May 1957, with the Roadster appearing the following year. To keep pace with the market, the XK150 had evolved into more of a grand touring car rather than a dedicated sports machine. Nevertheless, the 150's were very fast cars indeed with superb cornering ability and braking. In terms of overall performance, handling and sheer value for money, the XK150 was the best thing on four wheels.

In due course, there were a number of options available with the engine. Capacity was either 3.4 or 3.8 litres and ultimately both could be ordered in 'S' specification. The 3.4 came with the new 'B' head, which sported larger exhaust valves, increased angles on the inlet valves and twin SU HD6 carbs. In this form, the engine produced 220 bhp and was now as powerful as the XK140 with the 'C' type head. In addition, 215 lb/ft of torque was generated at 3000 rpm – some 1000 rpm lower than on the 140.

The four-speed all synchromesh gearbox was available with or without overdrive. A Borg Warner three-speed automatic with manual override was listed as an option, but was usually fitted as standard on 150s destined for the United States.

In spite of the increased weight of the XK150 models, the extra power produced a higher top speed of 125 mph (136 mph in the case of the 265 bhp 3.8 'S'). The conversion to servo-assisted disc brakes – pioneered by the racing Jaguars – was therefore a significant improvement.

With such power, the XK had virtually reached the limits of its performance and capabilities. Styling changes could not disguise the fact that the design dated back to 1949; the chassis was actually unchanged from the Mk V saloon and the body shape was no longer the last word in aerodynamic efficiency. The suspension, particularly at the rear with its dated springs and live axle, was at the end of its development. A new line of saloons was waiting in the wings, the styling of which was beginning to overlap the lines of the XK150. At the same time, the sensational E-type was about to assume the mantle as the premier Jaguar sports car. Consequently, in 1960 the XK range had reached the end of its illustrious life.

*As revealed by this XK150SE, the bodyline was much straighter than the 120's and 140's and did not have the pronounced dip in front of the rear wheel. Thanks to the slimmer doors, the interior was nearly five inches wider*

**Left**

*Initially the central panel of the dashboard was in aluminium, but this was soon changed to leather with foam and leather padding on top. Now without walnut, the dash no longer exuded the feeling of quality associated with the marque. This late-production car has a sliding heater switch under the rear-view mirror and an indicator stalk on the steering column. Curiously, there was a shortage of standard handbrakes for some of the later 150s, these having the American-type complete with a microswitch to show a brake light, requiring them to be fitted on the right-hand side of the gear lever*

**Above**

*The rear seat area carried over the improvements from the XK140. The seat-backs were detachable and could be placed at the side of the cabin so that an adult could sit sideways across the seat – a much more comfortable arrangement.*

**Left**

*The comprehensive tool kit even included a camshaft setting gauge. The boot was very similar to the 140's, with the spare wheel hidden away in a well beneath a plywood lid, which had the addition of a sheet of aluminium on the 150. The lid was protected by a Hardura mat. This prize winning restoration is by Brian Ekin*

**Above left**

*The rear bumper wrapped around the back of the car, the number plate being mounted on its own panel. Yet more chrome was used. This particular car is one of the later models fitted with triple light clusters and more centrally mounted overriders. As before, the bonnet and bootlid are in aluminium*

**Above right**

*The door trim included a combined armrest and door handle of substantial proportions. The quarterlight mechanism changed little for decades*

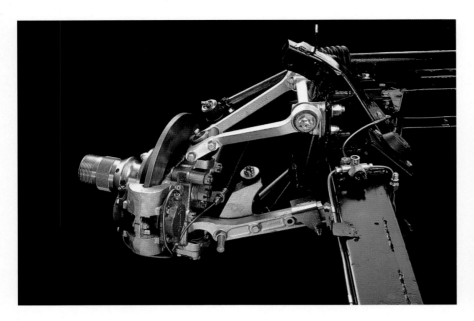

**Above**

*Dunlop disc brakes of 12 inch diameter with Lockheed servo assistance were fitted on all the 150s. Self-adjusting, the braking system required the minimum of pedal movement. The zinc plating is non-standard but is sometimes preferred due to its durability. Visible here are the grease nipples for the steering joint and the stub axle ball joints*

**Right**

*A careful look along the length of a car's body can be very informative. The bottom edge should be absolutely straight – it will show just how well the door has been hung and how accurately the panels have been fitted. Any reflections in the finish should be flat; if the various coats have not been properly applied, ripples will be revealed in the paintwork*

**Above**
*David Woodhouse is quite prepared to use his car in adverse conditions, even on a foggy day across the moors*

**Left**
*The wider, curved rear window gave improved visibility. The majority of 150's had the existing single rear light unit*

In spite of its more highly tuned engine, the 150 remained a quiet, effortless car with a wide speed range through the gears, top-gear being particularly flexible. Fuel consumption varied between 15-24 mpg, depending on the driving conditions and the driver's right foot

**Left**
*Show time, and last minute preparations and adjustments are made on this 150. Even though the radiator had a more improved 12 blade fan, it is quite common for owners to fit an extra one for better cooling, particularly for hot days in heavy traffic*

**Above**
*A large event can be attended by scores of XK's. By the time production ended in 1961, nearly as many Fixed Head Coupes had been sold as the other two models put together*

*When new the standard Drop Head Coupe cost £1793, rising to £2204 for the 3781
cc 'S' – the most expensive 150 model, but still outstanding value for money
compared to the arguably inferior competition from Mercedes and Ferrari*

*The Roadster joined the range in early 1958, shortly after the other two models. Fancy saloon gadgets like door handles appeared on the Roadster for the first time.. There were no rear seats but the wind-up windows were a distinct improvement on the previous separate side screens. The bulkhead was moved back a few inches, creating a longer front and bonnet. As the wings remained the same size, the doors were enlarged to compensate. Most 150's came with wire wheels and body colour spokes were an option. Roadster production amounted to 1297 (3442 cc & SE); 888 (3442 cc 'S'); 42 (3781 cc SE); and 36 (3781 cc 'S')*

**Above**

*Triple SU carburettors on the 3.8 'S' head, as restored by Patrick Lacey Restorations of Worksop, Nottinghamshire. The work of Harry Weslake, 'the straight port head' consisted of three twin induction pipes of equal length and was designed to enhance fuel flow through the venturis. With its triple carbs, stronger clutch and lightened flywheel, the 'S' engine developed up to 265 bhp at 5500 rpm; compression ratio was 9:1. Properly set-up, the XK150S was decidedly quick, 0-60 coming up in under eight seconds, while the standing quarter mile took only 16 seconds; top speed was 136 mph. Decidedly rare, only 36 examples of the 3.8S Roadster were built*

**Left**

*The steering wheel continued to be adjustable for length but not for rake. The dashboard on the Roadster retained the same layout as the other 150's, but was finished in vinyl rather than leather. As the Roadster remained rather draughty with the roof up, customers who required more comfort opted for the XK150 FHC*

# Restoration

The restoration of virtually any part of an XK is a specialist task requiring particular skills. Few non-professionals can do more than simply assemble components made and renovated by experienced craftsmen.

The XK is a complicated car and there are countless areas which can give problems. The many curves around the body make it susceptible to corrosion over the years. Matters were not helped by the fact that the original under-sealing was not oil resistant. The biggest and most expensive area is undoubtedly the body; it is not possible to be absolutely certain about the quality of some previous restorations without undertaking a complete strip-down.

**Above**
*Be warned. Stripping off the paint from this 120 revealed a mass of filler and disastrous metalwork*

**Right**
*A 120 and a pair of 140 Fixed Head Coupes all complete and awaiting restoration*

**Above**
*Serious surgery will be required on the worst areas and new sections welded in if new panels are not to be used*

**Right**
*All panels from the XK range are available as new components. Contour Autocraft of Spalding in Lincolnshire have a high reputation for quality workmanship and accuracy, having taken all of their patterns from original cars. The wings' complex curves make them particularly difficult to recreate – the edges have to be rolled and formed by hand*

**Above**

*A new piece is welded in place at the Patrick Lacey workshop*

**Above right**

*On the Jaguar production line, the light housings were welded on to bare metal and were prone to rusting. Now they are coated with liquid zinc before being spot-welded. A coating of 'Waxoyl' inside the wing gives extra protection*

**Right**

*Original Jaguar panels were accurately made to consistent standards, but the work of subsequent panel makers has often been of indifferent quality. Not surprisingly, some of the car's dimensions are inaccurate and much work is required to ensure that body components are properly matched. The craftsmen at J A Cook Coachworks of Nottingham needed to patch in a new section in order to get this rear wing to line up with the rest of the bodywork. Sorting out problems such as this takes time and can make a considerable difference to the cost of restoration. The rear wings are bolted on while the front wings are welded*

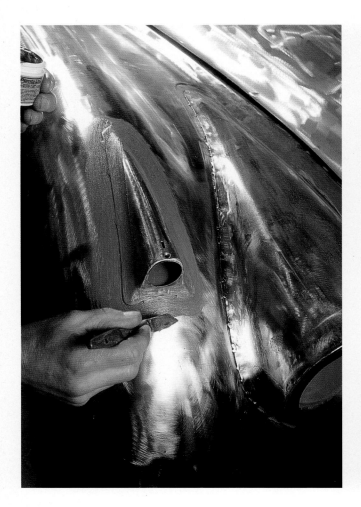

**Above**
*Lead-loading is an expression most people have heard about
but rarely seen. Once the new weld is completed, the joins and
edges need to be blended in and smoothed, and lead is used
for this purpose. Firstly, after thoroughly cleaning the area,
solder paste is brushed on and then heated to tin the surface,
thus giving adhesion to the lead*

**Above right**
*The lead contains some tin to make it more plastic. Unless the
working temperature is just right the lead will not have the
required fluidity. A leather pad or spatula is used to work the*

*lead in thoroughly. Tallow grease on the implement is
necessary to prevent it from sticking*

**Right**
*When cold, the surface is filed and sanded smooth. The area
must then be washed clean to remove any remaining flux*

**Above**

*XK140 doors had aluminium ends and steel skins over a wooden frame. The diagonal strut is for strength and to adjust the twist of the door as required for an ideal fit. The lower part of the door is prone to rust. It is wiser to apply a new door skin rather than trying to insert a patch, which will never look perfectly flat due to the heat distortion created by the necessary welding. Assembling the new body should start with the doors. After attaching the door to the hinge panel, the bulkhead may be adjusted to its correct angle. Spacers are used on the door hinges to give a straight door gap. When the door sill is in place the whole of this section should be accurate and true. The position of the catch and the door gap will, in turn, dictate the position of the shut face panel. The rest of the car can be built around these sections; with the aid of good panels, everything will line up correctly. Some restorers start from various ends and then try to make the doors fit later, causing all sorts of problems*

**Right**

*Many XK components are now being remanufactured and can be purchased new. But this is not the case with everything and attending auto jumbles is often the best way of finding that elusive part – and saving some money into the bargain*

**Above**
*A 'C' head in poor condition from a 120 – restoring a rare item such as this would be well worth the effort*

**Right**
*Once the metalwork has been properly repaired and renovated, it needs to be prepared for painting. The surface will contain pinholes and blemishes which, if left, will be revealed after painting. To correct this, a sliver of polyester filler is applied and then rubbed down with fine emery paper*

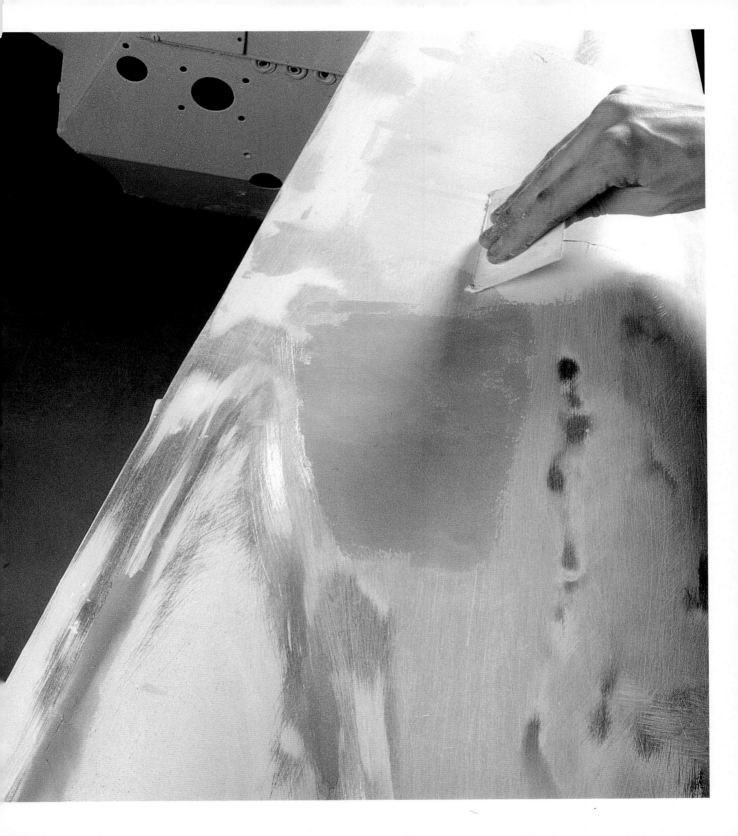

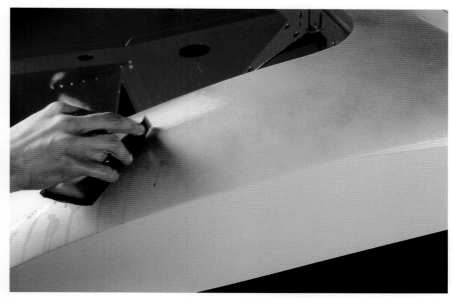

**Above**

*Between applications, a guide coat is used. This is a light mist of colour, so that during the flatting process, any high spots and dips will be revealed. It is important that flatting is done in all directions so as to ensure a flat surface without ripples, which would otherwise show up in the finish*

**Left**

*Painting is invariably done with two pack acrylics these days, because their durability is so much greater than the more traditional cellulose variety. Due to the content of isocyanates, however, spraying must only be done in a filtered and heated booth with the aid of breathing apparatus. A coating of polyester stopper is first sprayed over the lead, followed by an etch primer. After heat drying, it is flatted to key it before the application of three coats of primer*

**Above**

*Trimming is another specialist, skilled task. Kits are available for those brave enough to have a go themselves. There are kits to cover seats, all of the interior trimming and the hoods. Although appearing to be straightforward, there are numerous problems in obtaining a good finish. Aldridge Trimming of Wolverhampton have been established trimmers for well over half a century. Old seats frequently have to be restored with new backs and rubber filling as well as covers. After the cover has been secured, a steamer is past over the surface to tighten the leather and remove any small creases*

**Left**

*Some five to six top coats will be applied with heat drying after each application and flatting with fine wet and dry paper. After any specks are removed and bubble pricks touched up and blended in, the final coat can be polished and buffed. Undersealing is likely to comprise of an etch primer, primer, stone chip resistant paint and top coat*

# Specifications

## Production of Jaguar XK models

| | | | |
|---|---|---|---|
| XK120 OTS | *October 1948-54* | XK140 FHC | *October 1954-57* |
| XK120 OTS Special Edition | *August 1951* | XK140 DHC | *October 1954-57* |
| XK120 FHC | *March 1951-54* | XK140 automatic | *October 1956* |
| XK120 FHC Special Edition | *September 1952* | XK150 FHC | *May 1957-60* |
| XK120 DHC | *April 1953-54* | XK150 DHC | *May 1957-60* |
| XK140 OTS | *October 1954-57* | XK150 OTS | *May 1958-60* |

## Jaguar XK120 Engine Specification

| | | | |
|---|---|---|---|
| **Cylinders** | Six inline | **Valves** | Twin overhead camshafts |
| **Bore × stroke** | 83 × 106 mm | **Carburettors** | Twin SU |
| **Cubic Capacity** | 3444 cc | **Clutch** | Borg & Beck 10 inch single dry plate |

| Model | 120 | 140 | 140SE |
|---|---|---|---|
| *Compression ratio* | 7:1 | 8:1 | 8:1 |
| *Maximum bhp* | 60 @ 5000 rpm | 190 @ 5500 rpm | 210 @ 5750 rpm |
| *Maximum torque* | 195 lb/ft @ 2500 rpm | 203 lb/ft @ 3000 rpm | 213 lb/ft @ 4000 rpm |
| *Gear ratios* | | | |
| *Overdrive* | – | – | 3.19 |
| *4th* | 3.64 | 3.54 | 4.09 |
| *3rd* | 4.98 | 4.28 | 4.95 |
| *2nd* | 7.23 | 6.20 | 7.16 |
| *1st* | 12.29 | 10.55 | 12.40 |

| Model | 150 | 150S |
|---|---|---|
| *Cubic capacity* | 3442 cc | 3781 cc |
| *Compression ratio* | 8:1 | 9:1 |
| *Carburettors* | Twin SUs | Triple SUs, HD8 |
| *Maximum bhp* | 210 @ 5500 rpm | 265 @ 5500 rpm |
| *Maximum torque* | 215 lb/ft @ 3000 rpm | 240 lb/ft @ 4500 rpm |
| *Gear ratios* | | |
| *Overdrive* | – | 3.18 | 3.19 |
| *4th* | 3.54 | 4.09 | 4.09 |
| *3rd* | 4.54 | 5.24 | 4.95 |
| *2nd* | 6.58 | 7.60 | 7.16 |
| *1st* | 11.95 | 13.81 | 12.20 |

**Note:** gear ratios varied, particularly with US models